D1252259

Scientific and Mathematical Theories and Inventions

Noah Girouard King

(1) Quantum Energetic Theory:

It can be experimentally proven that, even on the molecular level, orientation is everything for any unbalanced or asymmetrical object moving in three dimensions, even when it comes to the exchange of forces. Asymmetrical, or non-spherical things that respond electromechanically to impactive shock across a perpendiclar axis of a weight and the speed, will transfer a different, lesser amount of force to whatsoever object it knocks into than one which is moving uniformly towards it in a parallel line. This is because the electrodynamical barrier between collision is capable of dispersing the force to other components, generating more heat and self-vibration on impact that does not necessarily transfer to the object impacted. Even without air resistance, a feather-type object that falls horizontally in a vacuum towards an object should theoretically bend and taper its force more effectively than a head-on colliding dart type object of the same weight and valence to the same object, because the electrons compressed by the collision have more free room to disperse friction and create other forms of energy in the initial horizontally diffusing scenario. This is important to materials science, and measures such as the relative surface tensions of materials and their abilities to support themselves over the surface tensions and become solutions on the molecular level.

(2) Physical Laws, Information and the Fundament of the Forces:

There are four types of material forces:
1) Geometric forces, which defines an ultimately repelling shape or boundary of influence by a functional or linear construction of ultimate directional quality, defining a particle of

definite occurrence by a semantic orientation of ultimate non transcendability by which it is not miscible or transformable by other shapes.

2) The geometrically proximal forces which repel, in a proximal, inversely proximal or instantaneous and unchanging manner, by a continuous addition or state of enactment of force from one particle to the next, but never bypassing or transcending the strength of the geometric force holding it unrelenting to its geometric form.

3) The force of the directional, the subtle force which defines movement in differing directions at times (which you may understand partially as inertial values): or, simply the statement of the enactment and instability, thus difference in force over the course of a dimensional direction, implying some inner property or force of the dimensional type may be discovered as having perpetuated a new dimensional phenomena of an unsymmetrical type, finalizing in symmetrical finalized measures of dimensions. The experience of rotation is the fundament of the variance in these dimensional/directional forces.

Additionally, it may be found in the fields of particle/elementary physics, as well as classical physics, to obtain a relativistic measure for any perspective of the balance of equivalent forces, by which particles are made from particles at large, that: In systems of particles where there is varying arrangements of a number of orientations or different permutations of possible combinations of types of particles, the minimal or maximally and most symmetrical group of substituent/composite particles may be used to provide an objective reference and charge for these particles, on any scale. Forces to which the minimal effect is achieved by the most symmetrical minimal/maximal construction of the coordinative series may be used to compare such relativistic measurements in more complexive multi-dimensional force/charge orientation matrices. In systems where composite particles and absolutely solid/geometrical particles are interacting as composites and no other comparatives are on that side of the complexity, then an ultimate relativistic measure can be made, but not assumed to be absolutely true, based on this type of spatial comparison. It may be predicted that the smallest point charge particle will either have an ultimately large and universally (simultaneously) forceful effect on the functions of these coordinations, or ultimately the smallest. This is the definition of the construction of the particle physics, and accurately provides analogy up to the level of conventional physics in such ways as chemistry, pH measures a prime example of the latter fundamental central-unit comparison. It may also be said that a cleanly oscillating system, not unstable or reactive with the media to much an extent, is typically required for emission or proper travel of the conventional sense for any particle system, and that the system becomes more 'clumped' and messy and non-instantaneous as the question becomes more similar to atomic chemistry and large-scale interweavings, unlike what is seen as pure light travelling in space. This is how certain particles which are not light may trigger photoreceptors of a given type, without ever really expressing a type of modulation, simply through an electric field adaptation. The non-linearity of summative physics is clearly astounding.

While a planet may be seen as whole from the standpoint of a heuristic boundary of the cross-distancial density of matter, the thing can not be seen viably as a definitive charge-holder in this realm of particle-based physics, for reasons such that the particles will arrange in any which way, still expressing nearly the same quality of force as per gravity, electrodynamic radiative summations, probabilistic and attractive measures, etc… on other particles in its field. A geometric definition and restriction of the particles to the effect that it begins changing its comparative function, as a whole, and definitely from a classificational point of view, to other particles of kinetic measure is what defines the difference by which I am talking. These are best seen as geometrically based occurrences, whereat large scale and nearly equivalent particulate orientations that are chaotic may be restricted to classical physical explanations, and matters of large scale chemical science (chemical equations being one important borderline mechanism of discernment of large scale classically physique systems).

I would assert that true particle physics is on the particulate level, but including levels such as the chemical. Whereas a divergence occurs, as any particular scale of physical classifications is reached, as the general heuristic boundary by certain relative (distance-wise oriented) scale variances and event-based analytical dissolutive-breakpoint heuristic constants vary while chaotic behavior erupts, but predictable and instantaneous matters of geometric nature fall to more abstract matters of a series, as of highly differing composites becoming abstractly and more specifically unpredictable as systemic componential transformations happen in our analyses (such as how we can compare humans and find little difference, but compare planets and see huge difference) to become invalidated and more increasingly complexly scattered and dissolute upon the effects of minute smaller particles: whereat we may conclude that the particle-based and higher physics will prove never to exceed geometric boundaries and simple displacements or breakdowns across time where units that are measured always behave as the units until the definite list of composites are rearranged as their identical constituents still but as enacted on a different system; while the predictions and proportions across what we expect to see from the scientific perspective no longer are effective when every object of a humanly discernable class now demands particular observance and respect to its individual functions. And so, essentially, because the energy of large scale systems to hold together are too weak and varied to become a small list of types of parts, it must be that it is no longer the same type of semantically operating system as smaller particle-based physical systems, since the variability among systems becomes dependent on a series of parts that are partially equivalent to the extent that they are removable from themselves; still non-interacting in proper classes (where displacements produce a lower change in time than those of other less analogous classes) for the most part yet changeable and essentially discernable as unique for a variety of kinetic factors in the system; and in other semantic bounds, without necessarily altering any other net charges to any opposite ends or dimensional directions such as gravity or basicity in particle-based physics.

Inertly oriented forces of action seem to hold true as simple dimensional values of relative movement across the spectrum of physical objects of varying definition, limited or changed only by other forces of any type or definition, though only through the quantifiably geometric and stationary systems such as humans in rockets/planes or photons on the particle-based level, or planets across the range of possible particulate systems. It may be observed that isotope-like amalgamations of yet smaller physical particles may express types of unpredictable statistical radiation emissions, but they are less common.

On another point regarding our familiar environment of electricity and atoms, one should have clearly noted that it can be asserted that spin-momentum of electrons should be represented as a product not only of the angle of impact and some random value of effective spin in correspondence to the directional influence, as it can be asserted also that the intensity of the spin varies upon the projected angle of interaction based on the direction and acceleration of the particles before and after the interaction. The impactive force of a charged particle upon it could vary as the product of the pressure and rotational velocity at the time of impaction and afterwards, the distance of the electron between two focusing materials in a molecular/atomic matrix, and otherwise. The repelling and focusing forces of nearby particles and the rotations/revolutions of the particles nearby, thusly, define more precisely the spin momentum, as well as polarity. The spin/polarity mechanics of photons and electrons may be seen through a particulate perspective, likely oriented in threes or sixes much like quark-gluons in nuclear matter (which, by the following logic, may be responsible in itself for a direct reference and correlation to the spin and polarity in ejected light), so as to provide the standard 3-dimensional electrodynamical perspective of dimensional movement values in all possible scenarios of travelling light, ordinarily as an exchange of force between three primary composite particles, oriented on a planar axis of geometric convention statistically representing polarity, and moving kinetically as a function of ultimate dimensional directional movement, and magnetically spin-responsive orientations of oscillation so that the weight of the particle is responsed to the oscillating rate, and the tendency of the electron to have a spin-state particle or particle grouping/proximity which tends itself to one point of the triangle's geometric phase, essentially suggesting a spin value which corresponds (as well geometrically) to the propagation at large of magnetic fields and the direction of an individual electron's polarity. I also theorize that positrons are equivalent to electrons, though their polarization of the phenomena in the 3-dimensional area is so that no polarity or polar momentum to the path is expressed, but rather an oscillation of the photon in a circular pattern around the path, and so polarity would not be expressed but an increased spin effect is occurred. And so it may be said, and as I theorize, that antimatter varieties of particles in known physics are just versions of the same particle, but tilted halfway around so that polarity or oscillations are expressed in a half-turned method across their path in comparison to the original, or are made of oppositely attracting sub-components when paths are not clear as in bosons. I theorize, through this, that the nuclear beta decay reactions creating positrons, and other decay reactions, rely on free photons, a free quark exchange(or just emission), and an extensive and instantaneously exposed boson field (see my later theories on bosons), forcing out an electron and simultaneously absorbing one nonlocally(most times) based on rest energy and polarization/direction of the electrons in the field of the boson embedded in the nucleon, especially so when the orientation of the photon/positron ejected in equilibrium with the decay and quark exchange is polar to the one attracted. Integrating this with our understanding of electrodynamics and materials science, we have also the following principle: Force can be viewed generally, including the physical materials science by a combination of forces of numerically interpreted potential to either push or pull a particular part over another, generally explaining electromechanical 'springyness' or kinetic deflection, and direct velocity transfer, and to provide interference curves of ballistic force transfer and breakpoints. Through this, I make a very interesting theoretical conclusion about the functions of electric force, and predict a theorem for the decay of an electron:

Oscillating electricity, upon reaching higher energy states, becomes more magnetically deflective by the forces of the interactions between the parts and the tiny particles I theorize conjointedly to be holding together the electron in analogy to the nuclear quarks and gluons, in a way whereat it becomes vibrating across itself in an oscillatory fashion with a direction, so as to be less like a particle electron and more like a photon, but only until a particular point - then it may at this point dissolve and dissociate with other electrons or even other photons, providing nuclear stability: this point, at which it is possible as per the energy of the inner workings of the light that the particles dissolve as a result of the loss of adherent and cyclic balance of the repelling, attracting, and stretching energy of the different parts of the electron in effort to apply mathematically a limit at a certain thereof calculable energy to the particle upon finding the constituent parts by assuming there is a finitely surpassable boundary of the substituent particles, as per the difference of their attractive forces and the force pushing them apart in the process of generating the oscillating energy, which also explains the size of the photon field becoming slightly larger upon ejection at higher energies, but more pliable magnetically. It's possible, in chaotic and energetic scenarios where the size of possible atoms is too large such as stars, that the decomposition and reassembly of light may become recognized as a common dissociative phenomena of nonlocal rearrangement of electric matter during stabilization. This mechanism distinguishes much less the photon from the electron, but describes a three-part energy phase correspondent to nuclear matter.

I go on and theorize that if the fundamental components to the electrical-quark field interactions, the Higgs boson particle, or rather the particle responsible for a theoretical field propagating the interactions between electrons and nuclear matter, were to be put inside out in its internal structure, jumbled around a bit, stereoisomerized in its elementary connecting structure or provided a simple phase change, the field it generates which theoretically determines electric and nuclear interactions should follow to that all its determinate influences would be probabilistically inverted to produce transmissions or reflections at absorption, spin change, absorption or transmission at reflection, or reflection and absorption at transmission, or some mix and match operation in the localized field at the level of emission and absorption to some degree of intensity. Perhaps the charges would be reversed in a case. Perhaps a change of time vectors in these behaviors, or a permutation of elemental matter's schematic would shift the structure of the universe, put all the socks inside out. Things would look very much different under these effects, if only on a small quantum scale. I presumes this particle, as a possible reference to consciousness theory as it may condense in genetic information's exponentially high bond energy, may not be a boson-group particle but the only particle/anti-particle to its own class, as I also would propose that this particle isolates itself into electrical fields, propagating one or more electrons, and feeding off an electric field (rather than attracting and then focusing, more weakly and in quarternary methods of cross-interaction with quarks in general like bosons do), into electric fields.

(3) Objective Observations:

Gravity and Relativity:

An observational device may be constructed out of minute quantum or elementary microscopes, and be tracked in data with a gravitometer and satellite technology to provide perfect technological reference to areas in space and relative velocity versus the relativistic location. Using these values, one can prove exact relativistic constants, despite gravity, and begin to predict how gravity effects photoelectricity versus other matter. The actual definitions or explorations of deep space are incalculable without these relativistic compensators.

Observation:

Minute material particles can be measured by bombarding the particle with other smaller non-invasive particles, and exposed through tachyons or interference to electromagnetic sensors, in order to be measured directly over a large exposure period, or more trickily by an exposure diagram in interference with a (only preferably) larger field particle. This is the key to solving the structure of matter.

Batteries:

Some type of battery mechanism that recycles halides to Li+ or Ni+ ions, stores energy to reheat and recycle the spent ones, through nanotubules in water, is the key to municipal chlorine recycling and osmosis filtration.

Electrochemical Stitchwork:

Chemicals which present resistance of a type to electric conductance, and are not normally prepared to react, can be placed to a compensating electrode/electrode environment or housing/plating to provide for alpha-position reactions in the properly ionizing media where the ionization would not normally present itself, along with further reactions at the proper energy levels.

Assuming materials are soluble in a liquid as a product of both of their electronic energy levels and phenomena of structure, it can be assumed that a solution boiling at a particular rate can be relied on to produce a certain amount of residue of solute material that is dependent on the level of solubility of the solute up to pre-boiling temps, and a reduced or greater amount of solute normally soluble in the medium would turn out likewise or conversely as it transitions towards and into the solid state. The complex curves and tendencies of the different states, and rate of liquid/solid/gaseous state matters' solubility curves can be indexed and understood experimentally by the relative measures of separate residues as they are but then be compared as a factor of residue or dissolved weight in gaseous medium in the atmosphere or in a vacuum, and observations about the electrical contortions of the materials in different phases can be inferred through this experimental set of measures for the function of the industrial via phase-change extractive residual expectation guidelines, and an analysis of the time to evaporate/evacuate versus product loss can be understood to optimize efficiency. Likewise, solvents are electrochemical modifiers in that, for the temperature and structure, they alter chemical nature in the following way: Solvents and co-dissolved materials effect each other in the rate of reaction and equillibrium levels of different chemical reactions, and so can buffer or catalyze reactions as is well-known. This effect is also responsible for a change in freeze and vaporization temps, through electrical attraction and electrical conductance/capacitant properties.

The cyclicity of a molecule's electron orbitals and placement in tensor fields over time may be simplified by a theorem simply involving the size and number of the valent orbitals, the attraction of the nucleus as a factor, versus the other paths possible in the orbital and accounting for the probability of the spin/orientation of the flow in the field to more easily predict the positions, computational and syncopative tendencies of parts of a molecule versus another. This allows for computational variance and subtle molecular kinetic properties to be represented by interference, and can symbolically translate to represent the wave-mechanical cycles of conductance amplitude generated by certain molecules at different parts. This is simpler than calculating a tensor field, and may help to explain orbitals within atoms as well in the minor case.

Another interesting electronic/atomic phenomenon is the one based upon my noted assumptions about the trinary structure of light and frequency/polarity substituent conformities, and it declares the phenomena or principle of the connectedness of matter to singularity: by the rule that low-frequency light which has a more linear and thin range of dimensional oscillation, is the kind less likely to transfer oddly in the general circumstance to atoms; higher frequency radiation directed at any point has a higher vibrational tendency, thus further upsetting the perpendicular flow of the colliding electric fields by that tensor-based proportion of action; effectively fundamental to, but not in entirety, such abstract phenomena as the listed: the phenomenon of angles of incidence in reflective surfaces and diffractive indices. Ideally, this means that infinitely low frequencies have the highest probability of either passing straight through a substance, or reflecting directly back.

(4) A Proposal on the Oscillation and Movement of Shapes, and their General Mapping:

It is shown in the interfering structures of geometrical figures of the 2nd dimensional order of a monadic/circular construction, that: various shapes whose center point exists at the origin of a graph, when the two solution values for the values of distance across any particular axis are added and placed back uniquely on a plane and possibly rotated, that:

1) only the even numbers are capable of producing a flat signal.

2) triangles can produce a perfect triangle wave.

3) further rotations and segmentations create increasingly eccentrically sloped combinations, up to pseudo sawtooth formations, of many subcycles.

(5) Power, Kinetic and Conservational Technology

I propose for and recommend a design for a gas-pressure turbine / gaseous combustion engine as potential, which would use mixture of platinum-lead alloy electrode at 16.5% lead to platinum by molar quantity, this alloy to be called Elucidium, aqueous mixtures to produce an electrolytic catalyst towards a reasonably suitable fuel source that could theoretically sustain enough production of combustion or pressure energy upon electrolysis that the engine which is turned by the hydrogen or gas pressure produced can additionally power the continuance of the power cycle while sustaining rotation of the shaft. I otherwise suggest catalytic electrolysis be used in the reduction of gaseous outputs from motors within the apparatus.

(6) Semiconductors and Circuit Elements:

I propose the following semiconductor or other involved circuit elements for general purpose be prescribed to be researched and theoretically implemented in computers and elsewhere immediately:

One of my theories based on laws of conductance and my own electrokinetics and electron mobility calculations entails that it is advisable that an alloy of 16.5% boron to 83.5% molar silicon, can be employed as a single unit of semiconductor, as it would only cease to strongly inhibit the transfer of current at a sharp breakthrough point, and the proportion of surface area of contacts can generate the effect of amperage in a direction on the circuit. The alloy of 16.5% boron to 83.5% silicon may be called Evaluatum, for its intended use of being a simple alloy for evaluating electronic logic expressions, predictably this is more likely on a larger scale where my calculations suggest it could be tested.

Biodes, a one-input and two-output semiconductor diode, process and simplify transistor processing fundamentally in computer science, by halving voltage and recomparing dual voltages for re-input to the computer apparatus, as well as splitting voltages in power supply units and memory, in conjunction with the Transratiometer.

Transratiometers, a two-input two-output semiconductor barrier, follow a mathematical principle where a parallel conductance is sustained, to be filtered to a proportionate current based on the proportion of distances and voltage from the two inputs to outputs, upon base electrification. At the base proportion of one half current to each output without bias, the transratiometer proves itself highly useful in computer science, as a replacement to many transistors in certain scenarios, as well. The transratiometer may also not express ratios between the two outputs

Limistors simplify the a/c adapter, transformer and more, plus serve as filters and power splitters, for high proportionate voltage splitting.

Two output or transistent transcapacitors function as dielectrics, but based on the levels of attraction or deactivation of the dielectric molecule medium to different complex semiconductor materials (needs research)

Of the Biode and Transratiometer: these new implementations can actually complete their own form of logic gating in computers and do a variety of signal manipulations for more complex computing, such as checking two or more AND values successively, or supplying more OR gates for a number without as much space as transistors alone do, at an exponentially more efficient rate. Theoretically, because of how they can simplify processes in computers, it is possible biode and transratiometer may greatly reduce the size and wattage of most computer chips. Examining the effectiveness of the new components is a matter of looking at the designs for a series of computer chips, and then writing schematics that implement the biode and transratiometer for an equivalent chip. The equivalently functioning chip should express multiple benefits over the old model, and be much simpler to make. I predict that computer chips will be up to 25% reduced size, and 150% or better comparative power consumption rates and speed on average. What follows is a series of diagrammes proving for the biode and transratiometer effectiveness.

Key

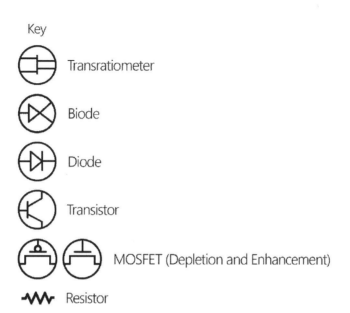

Transratiometer

Biode

Diode

Transistor

MOSFET (Depletion and Enhancement)

Resistor

AND Gates

Fig. 1

Fig. 2

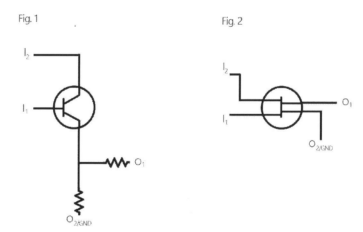

Fig. 1 Represents a traditional AND gate, a total of 3
components. Fig. 2 Represents an AND gate utilizing
the Transratiometer component, for 1 component in
total. The first circuit has a higher current than
I(2) or I(1) for output, necessitating the resistor,
the second circuit has a mean of the two current
inputs as outputs.

OR Gates

Fig. 3 Fig. 4

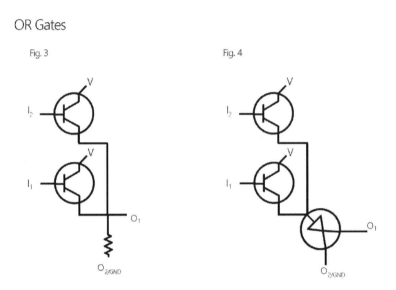

Fig. 3 Represents a tradtional OR gate, using a
resistor due to excess wattage at the output. Fig. 4
Represents a similar OR gate using a biode component,
reducing overall wattage.

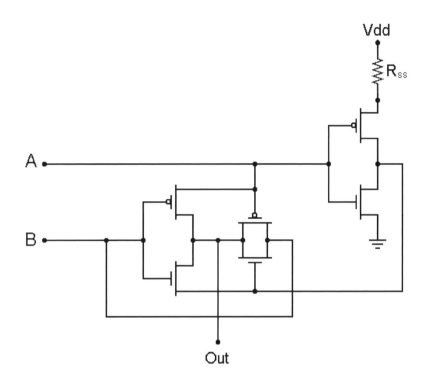

[https://en.wikipedia.org/wiki/XOR_gate] Image depicting an XOR gate, for a total of 7 components.

XOR GATES

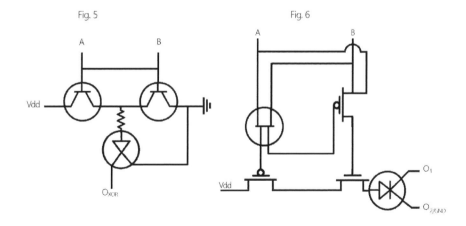

Fig. 5 and Fig. 6 represent XOR Gates that use less components and perform virtually the same function, Fig. 5 using a Biode, and Fig. 6 using a Transratiometer.

NOT Gates

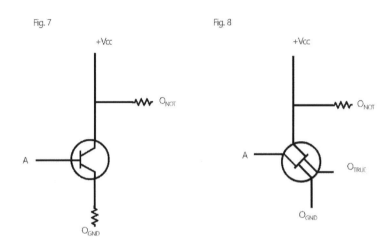

Fig. 7

Fig. 8

.

Fig.6 Represents a traditional NOT gate. Fig.7 Represents a NOT gate using a transratiometer, which can also be used to supply an additional TRUE signal, and does not require an additional resistor to ground.

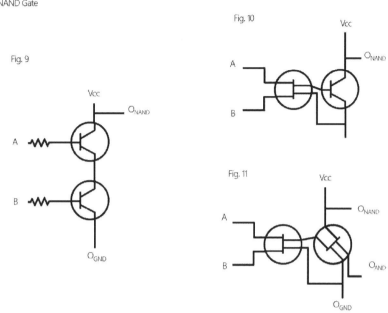

NAND Gate

Fig. 9

Fig. 10

Fig. 11

Fig. 8 Represents a traditional NAND gate. Fig. 9
Represents a NAND gate using a transratiometer, which
splits the total value of the combined inputs A and B,
removing the need for resistors. Fig. 10 represents a
similar circuit where the transistor component is
replaced with a transratiometer, allowing the gate to
save an additional AND value.

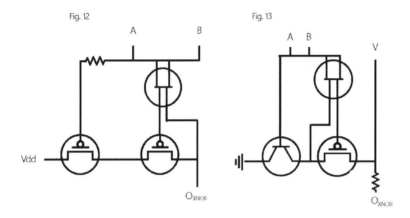

Fig. 12 and Fig. 13 represent 2 XNOR gates, showing how transratiometers can be used in different ways as an AND signal component to solve a circuit.

Biode

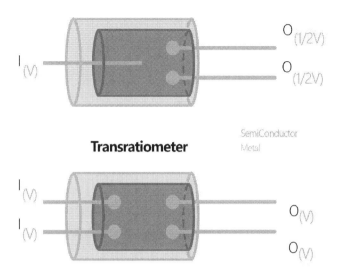

$O_{(1/2V)}$

$O_{(1/2V)}$

$I_{(V)}$

SemiConductor
Metal

Transratiometer

$I_{(V)}$

$I_{(V)}$

$O_{(V)}$

$O_{(V)}$

BIODES AND TRANSRATIOMETERS

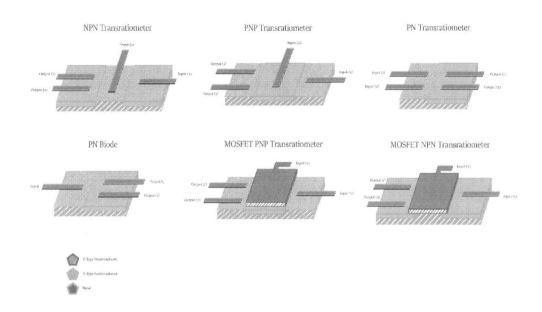

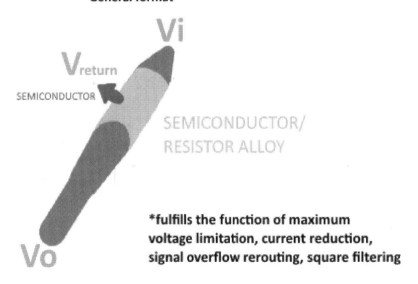

Limistor
General format

Vi

V return

SEMICONDUCTOR

SEMICONDUCTOR/
RESISTOR ALLOY

Vo

*fulfills the function of maximum
voltage limitation, current reduction,
signal overflow rerouting, square filtering

TRANSCAPACITOR

a capacitor that has two inputs,
one which aligns the dielectric
and another stationery charge
applier,which can reroute the
current.

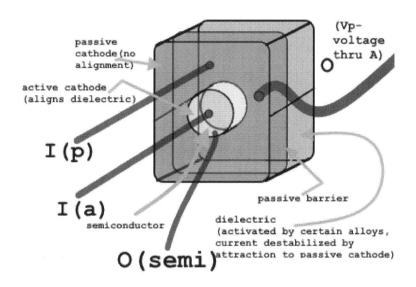

(7) Minor Physical Forces:

An optical effect of non-correspondence
probability offset in reflecting and emitting light
can be attributed to weak nuclear forces responding
asymmetrically or simultaneously to other factors
(including other interacting photons/electrons) while
the optical or electric event is happening. This has a
minor effect on polarity, spin, and reference angles.
Through this effect, it is most probable that even the
most theoretically predictable effects may still
harbor unpredictable statistics through the nuclear
interactions.

(8) Particle Physics Theories:

Elementary particles might theoretically be emitted from and observed in matter in the following way: Electrical and nuclear stimulation to atoms or other materials in a geometrically isolating fashion (where there are boundaries and a sample of atoms tested) may cause them to emit corresponding smaller subatomic/elementary particles that can be differentiated by mixtures of the materials and the types and nature of the energy present. Their particular affinities, these different materials, can be leveraged to obtain more types of particles to potentially even perform advanced alchemy. Elementary particles, I theorize, interact with their environment in particular field quantas, and amalgamate themselves at centers of fields to which they are individually connected. I predict different quarks function in reference to each other, by the sense that they separate to different polarities in electric fields, and that they singularly may appear minutely within atoms of large nuclei, helping sometimes to maintain nuclear valence and informatics or internally undisturbed resonance structures. Certain types may be more common in isotopes. Therefore, a variety of the theoretical quark and boson emitters should be produced and tested.

Likewise, performing particle acceleration experiments on certain materials is bound to give coordinational sequential reference to amplitudes at random to emit certain types of particles.

(9) Mathematics, computers, and geometry:

Wave Constructs:

One can mathematically determine the average level of combined wave phenomena in a harmonics construction by indexing segmentations of oscillation that exceed a limit value, versus the length over a period that falls under the limit value (for example, 1/2). Alternately, a series of waves can be measured for their periodicities and total amplitudes, so that a phase of each wave individually at their different major convergences and divergences can be added along their syncopative cyclic interval, permutatively, to show the satisfaction of the limit value in proportion to the wave versus the maximum and minimum, demonstrating the capacities of the waves to trigger phenomena or other mathematical values. The waves can also be analyzed of the probability of intersection of the frequencies at different phases by the comparison of the frequency ratios, and through permutative multiplication they can be used to determine other summed amplitudes of occurrence or satisfaction and be reconsidered individually for their probability to satisfy the limit based on the proportion of the limit occurrence, showing comparative functional reinforcement or negative interference; otherwise the sum may be measured and deconstructed individually in relative limit-comparisons to achieve a similar answer. The frequencies, when added, should reduce to a low proportion in the most predictable and instantaneous positive interference setups. Solutions can be found where the wavefunction finds a solution for 1/2 the maximum amplitude, and relative measures of this versus the average amplitude can determine the factor of relative harmonic directionality, in terms of the complexive positive/negative interference rates.

An interesting, new way of processing fractal data not limited to single solution point analyses such as of the Mandel or Julia sets: graphs via comparative analysis heuristics for solutions based on finite, complex dimensional number set coordinations which can be measured on their own based on self-sameness and terminating/exteriorized angular coordinations of their own/against a containing set of dimensional measure, or to be compared based on conjunctions of the values and positions of previous equivalent solutions. Such heuristics should represent fractallescent amplitudes of such things as multi-step energetic reactions stemming from symmetrically dispersing matter.

Number Sequences:

A new number counting sequence, to be called the summative branching sequence, can be indexed by the following heuristic: Each repetition of the heuristic, without preserving the count of branches from earlier generations, may each on their own refer to n+1 iterations of multiplicative branching per unit stem. The sequence goes as follows from n=0 to n=6: 0,1,2,6,24,120,720... It may be noted from n=0 -> n=2, the number is equal to the number of the previous of sequence, plus one. From n=4 -> n=5, the pattern is of the number before it in the sequence, multiplied by three then four. This is a novel alternative to Fibonacci sequencing, and may appear often in internalized, branching mechanisms such as neuron cultures.

Computers and Geometry:

There is a number for each cyclic bit interval of cyclic binary counting where the number of times a 1 is added to the binary value reach a change in the value for each part of that number follows a curve approximate that to a circle, before shearing off exponentially. Thusly, the bit order is geometrically imitating a virtual sine curve in the sense of chronological measurement of parallel augmentation upon adding incrementally. Also, in analogy to the fact that 3-dimensional solids follow the limitation that the highest unitary compositive face type is a hexagon without other numbered shapes' involvement, there is a bit-per-number size that, when defined with a number approximating the previous digitally abstracted hemicircle, corresponds in its add-to-change capacities to the approximate values of a circle or sine of a certain size, without shearing off so clearly as exponentially. This can be referred to as a perfect probabilistic hemicircle of computer probability, the exact number I have not calculated. The number of bits to satisfy this is expected to be a low number, analogous to the number six being the highest 2-dimensional solid to tessellate into 3-dimensional solids by themselves. Some types of combinatorial computational processes may be restricted to this data size, perhaps by annihilating geometrically the bit values, to achieve a certain type of unique and novel answer. It is interesting that there are calculable cross-references of possibly inferred mathematical sloping in these binary number sequences.

An important new fundamental construct of computer bitwise operations which results in the occurrence of multiple bit value states, requiring free valence neutrons in the inner shell in such a fashion that they interfere upon each other against a primary comparative value upon two, opposite, of six electronic contacts on either side of the nucleus, increasing the valence neutrons' impact on the collector contact, in a fashion so that higher orbitals of transfer electricity are then excused from the site of neutron disturbance, and their spins are deterministic of the polarity of the electron transfer points of the conductor. The probabilities are verifiable in a stream of nine major sections, a 3x3 square of angular deflection by the comparative influences, and also by the level of the voltage provided in terms of the balance of the projection and the voltage required to produce it. This leaves two values each for 4 spaces, and 3 values for the middle. So, there would ideally be up to 15 values generated by such a processor mechanism, and from just 6 valence holes in a quantum computing scenario.

Here are a set of equations that cover the phenomenon of the propagation of shapes, counted in order of the number of vertices, as they extend to infinity, and their volumes as they increase in faces within a circle, or by faces of unit length, also a list of equations for the solution of 3d solids that are tesselatory, or non-tesselatory and broken into other n-faced shapes by their measured averages of face numeric unit with all included faces at corners, together. This accounts for volume, radii, etc. The equations are:

2D Shapes Equations

Monadic Shapes (within unit circle, radii = 1)	Regular Shapes (outer segments = 1)
Radii: 1	Radii: $(\frac{1}{2})csc\left(\frac{\pi}{n}\right)$
Inradii: $\dfrac{cos\left(\frac{\pi}{n}\right)}{2}$	Inradii: $(\frac{1}{2})cot\dfrac{\left(\frac{\pi}{n}\right)}{2}$

Area: $(inradii * vertice\ length * n)/2$	Area: $(inradii * vertice\ le\mathllap{}$
Circle Area: $\pi(1)^2$	Circle Area: $\pi * radii^2$
Vertice Length: $\sqrt{(\sin\left(\frac{4\pi}{n}\right) - 0)^2 + (\cos\left(\frac{4\pi}{n}\right) - 1)^2}$	Vertice Length: 1

3D Solids Equations

Solid Radius(f)	$\csc\left(\dfrac{\pi}{N[f]}\right) * Radii[f]$
Face Angle(f)	$2\left(\arctan\left(\dfrac{\dfrac{Radii(or\ avg.\ inradii - radii}{for\ odd\ verticed\ shapes)}[f]}{height[center\ f\ from\ (0,0,0)]}\right)\right)$
Inface Angle(f)	$2\left(\arctan\left(\dfrac{\dfrac{Inradii(or\ avg.\ inradii - radii}{for\ odd\ verticed\ shapes)}[f]}{height}\right)\right)$
Side Angle(f)	$4\left(\arcsin\left(\dfrac{\frac{1}{2}Segment\ Length[f]}{solid\ radius}\right)\right)$
Occupied Arc Area(f)	$N[f] * (\left(\dfrac{1}{2}\right)$ $* (\left(\dfrac{\frac{inface\ angle}{2} * \pi}{* diameter[2 * solid\ radius]}\right)$ $\left(\dfrac{side\ angle}{2} * \pi * diameter\right)))$
Volume Average(f)	$\left(\dfrac{1}{3}\right)Area[f] * height$
Total	$N[f] * volume$

Volume of Solid	
Total Cover of Solid (should equal a whole #)	$\dfrac{Occupied\ Arc}{(4 * \pi * solid\ radius)}$
Surface Area of Solid	$N * Area[f]$

A series of Fractal Equations were also discovered by myself. The fractals exhibit various qualities and geometric tendencies. The Divergent Space set has few or no solution that is rationally calculable, but the lines can be approximated in their location by point closeness to a solution area. Most other fractals have clear solution areas. The fractals, for all values on the directional dimensional axis (Y) and for all alternate dimensions cross-factored (X) with complex number indice of distance of the point selected from the solution point (C) and their aptly chosen names are on the following table:

Fractal Name	Fractal Equation
Divergent Space Set	Y = xc + c
Wormhole Set	Y = x^c + x
Wormhill Set	Y = x + c
Worms Set	Y = c*(root(x))
Star Set	Y = c^x
Starburst Set	Y = (root(c)) * x
Supernova Set	Y = c^x
King Set	Y = x^c

Medusa Set	$Y = (root(x*c))$
Sunrise Set	$Y = x^{(root(c))}$
Meteor Set	$Y = (n>1)^c$
Meteorite Set	$Y = (n>1)^{(c*x)}$

(10) Genetics:

Synthesis of genetic material or protein using defined spectra of biological or biologically related materials in the presence of predictable electrocatalysts, through electric arc discharge in tunneling apparatus with a varying voltage, a homogenous mixture of carbon and other genetically derived (experimentally, as from % atomic ion by mass of standard sample genetic materials or substituting total cell nMR mass) material of a particular sinusoidal (as predicted, or otherwise mathematically constructed) gradient may produce complex organic material which is predictably and from a biological/ecological standpoint corresponding, of the depth of conductant layer from the cathode to the anode, and which will provide for an exact sequencing, which can be measured for concentrations on a qubit-wise (or, via the generated molecular formulaic signature) scale in order to find out the functional group of proteins, genetic sequencing, etc. available at such a voltage and at such a part of the apparatus. Desired parts, through filtering may be obtained. This is a guaranteed way, under laboratorial, experimental and medical/bioengineering circumstances, to create replicates of relevant RNA and DNA information for human or other use, through a delicate set of electrically monitored reactions.

(11) Complex Axioms:

A sphere containing any number of equal surficial

subdivisions of informational quanta can be
coordinated and arranged on an (n-1)-dimension
coordinate, according to the differences in its
objective location/distance, with the nth dimensional
coordinate multiplying and added to its directionally
relative amplitude (or simply made separate by
distance in each of n-1 direction and compounded as to
be of higher amplitude by/with the total n-dimensional
distance, or compounded along that n-dimensional
distance outward), to represent all the angular/scalar
datum in an equivalent metrical permutation which can
be seen on the lower dimensional scale as an effective
representation of the n-dimensional datum in its
combination, when referenced at an origin angle, of
the particular answer and coordination at which it
lies. Realize this is a useful, model for
datastructure indexing in processing for certain
applications, and also a good model for creating fun
puzzles and phase-based amplification and strange
optical effects. Upon the consideration for the
orientation of multiple instances of these
referential, listing, consequent objects with their
own complexly rotational centers or a unifying
rotational center, the exploration then finds us with
the following axiom: a 180 degree rotation of two
equivalent objects where one refers and relates to the
other in a certain way will reverse the order of
relation in its semantic value in certain measures,
and the previous. Likewise, a 180 degree rotation of
one or more different objects as a whole where they
and their parts refer to another in a certain
predictable and symmetrical way, will cause the
reversal of the order of relation in its semantic
meaning externally for the previously indexed
measures. If the semantically compared objects are
referring against each other equally, the reference is
simply reversed but the equality is preserved: though

this is not true if there is a directional-sensitive barrier — the order would be reversed, and so the mechanical effects expected would have to be considered for in the reverse behavior. This, however, is still always true for atomic samples, or any time for where each semantic part of the system is on a one-dimensional plot of behavior predictability. Also, between two parts of complexly unfolded n-1 dimensional arrays, when combined and unfolded across any origin coordinate after a 180 degree rotation of both individually, can be graphically analyzed in any other directional coordinate to find that the products of the arithmetic or multiplicative/divisive conglomerations in the same direction will prove to have the same (in multiplication) or inverse (in division) proportions to their values, as well as the differences being equidistant but polar to the first value than before (save for the difference of the first and last number) and the sums remaining the same. The multiplicative is the only preserved value when comparing the two despite the rotation in either transformation.

(12) Various Instruments:

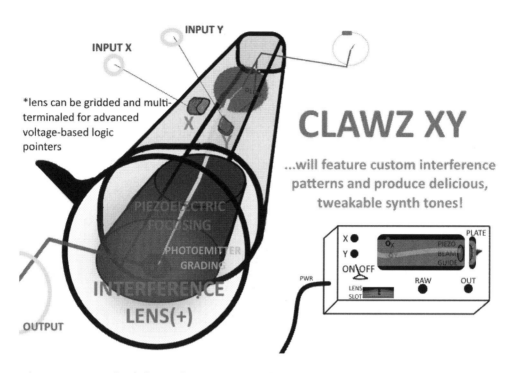

CLAWZ XY

...will feature custom interference patterns and produce delicious, tweakable synth tones!

INPUT X

INPUT Y

*lens can be gridded and multi-terminaled for advanced voltage-based logic pointers

PIEZOELECTRIC FOCUSING

PHOTOEMITTER GRADING

INTERFERENCE LENS(+)

OUTPUT

X

Y

ON\OFF

PWR

LENS SLOT

PIEZO BEAM GUIDE

PLATE

RAW

OUT

The general idea for an analogue of the Clawz XY is for a closed circuit, ending and grounding past a variable frequency or multiple colored diode in a chamber controlled and filtered in amplitude and frequency by two inputs, either solid signal or of audiomodulated levels, focusing into a beam to parse through a prism, farther down the chamber. The frequency of the light being emitted changes the path by which the light projects through the chamber. If the light contacts a plate at the other end, attached to a knob on the exterior of the chamber, a grade of photoconductor on the plate powers a circuit to audio output, after some amplification. All eight of the knobbed plates can be introduced, taking up 1/8th of the total space in the range of diffraction from the diode in the glass prism, to transfer signal. Otherwise, they can be set aside in a far part of the chamber. All kinds of wave mechanical shifting effects and pulse waveforms can be adjusted on these plates. A rubber barrier and vent hatches allow for easy dusting, and solid signal clarity.

(13) Synthesis and Extractions:

Neuromelanin, and Adrenochrome, are suggested to be able to be made according to the information on the following graphic:

*AgCl is recommended for conversion to Neuromelanin to synthesize Adrenochrome this way.

An analogue of Neuromelanin, an important brain pigmentation and potential supplement, can be made from deprotonated L-Lysine and Arginine in a 1:1 solution, with deprotonation at 6 sites. The two molecules actively react at these sites in each other's presence with the deprotonator (or Lewis acid such as $MgCl_2$)without extra catalysis to produce a Neuromelanin analogue as well as gaseous wastes. HOOH, hydrogen peroxide, is an equivalent deprotonator that helps complete byproduct elimination, and synthesizes water in the presence of the amino acids, leaving a cleaner synthesis.

Rich soil can be mined by filtering melanins through it, yielding mostly pure precious metal in melanin solution.

Precious Metals can be mined from rich soil using melanins, water, and heat. The following pictures describe my procedure for extracting non-magnetic metals (silver, gold, platinum) from ground soil, yielding over 1oz metal from just one bucket of soil:

1. Raw soil dig

2. Bucket with melanin water added

3. Melanin soil straining

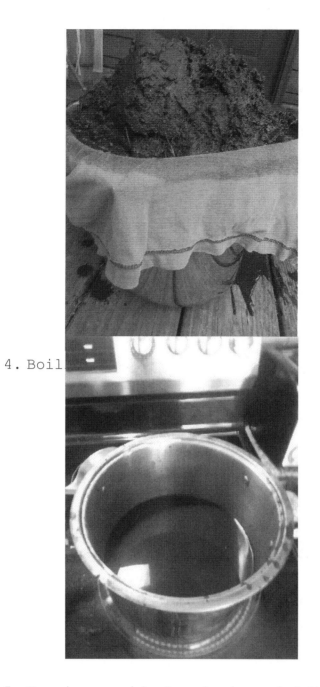

4. Boil

5. Burning residual melanin, yielding gold.

(14) Medical Science:

Injections can now be delivered from their medium in a spray form at a high pressure to break through skin barriers; to a spaced and padded applicator region from a high pressure jet on an applicator device designed to be pressed against skin. This is effective for telomerase dosage, for example, to be alternative from stomach breakdown and to reinforce the skin chromosomes. Wrinkles would be removed, and the chromosomes would absorb the telomerase better this way, replacing botox-like therapies. In general, this is an effective form of extended-release drug application therapy.

(15) Superconductors:

I have discovered a minor amount of resistence to movement, regardless of the field vector, of magnets and objects of magnetic character brought about by gold, platinum and silver atoms in a matrix, the matrix of the previously described gold/silver/platinum extracted from soil. As these metals exhibit or promote unit or marginal on an electron less of electron valence nonisometry, they project randomly magnetic fields, creating an occupational resistance of movement in any direction while locked in a magnetic field. It may be possible to energize such a solid matrix to produce a quantum locking magnetic field.

(16) Numbers of Interest:

$$\left(\frac{x}{x-1}\right) - x^{\frac{1}{x}} = 1$$

This formula may pertain to its single root as a number whose property allows it to be able to be used to predict the complexity of closed systems which sustain with maximum efficiency, or the order of multiplicity which things can be measured most effectively. Moreover the value of the roots of this function may provide a view into the maximum complexity of an arrangement of things whereat the most complicated sample of possible things, such as another curvature shape or equation, is required to be used to fully explain the concept of such a thing. The value has not fully been calculated here, but the number this formula is rooted at, is so far 1.72… or so.

Another number of interest is equal to:

$$I = 134279985$$

Because it is equal to an ideal and comfortable expansion of the values between two and three, the first two integer numbers to alter each

other in a series of multiplication that are
not identical, in consecutive iteration towards
an increasing increase in value by mathematical
operation:

$$2*3 + 2^{3+1} + 3^{2+1} + ((3+1)^{2+2} * (2+1)^{3+2}) + (2*3+2)^{(3*2+3)} \dots$$

$$\dots = I = 134279985.$$

I can be corresponded in its construction to a
series expression with incrementing base
values:

$$\sum_{n=0}^{\infty} \frac{n}{(n+1)*(n+2) + ((n+1)^{(n+3)} + (n+2)^{(n+2)}) + ((n+2)^{(n+4)} * (n+3)^{(n+3)}) + ((n+1)*(n+2)+2)^{((n+1)*(n+2)+3)}}$$

This value product of the previous
series, with its inverse defined as
L,

L is $= 134279984.768$.

And let K $=134279984.884..$ and
is equal to the inverse of...

$$\sum_{n=1}^{\infty} \frac{1}{(n+1)*(n+2) + ((n+1)^{(n+3)} + (n+2)^{(n+2)}) + ((n+2)^{(n+4)} * (n+3)^{(n+3)}) + ((n+1)*(n+2)+2)^{((n+1)*(n+2)+3)}}$$

While:

$$\frac{I - L}{I - K} = 2$$

Astoundingly you can see that this number has a plethora of incredible properties.

So, you can say that number I and L and K share some astounding trends to do with their construction and the nature of their relevant expressions which terms itself in a way that is similar to how algebraic statements will begin to approximate towards true equality in numerous formulas as their extends towards infinity. An example is how the sum of inverses of a number to n power becomes closer to the inverse of that number minus one. Potentially this number I can be used to replace true infinity in certain calculation series in applied mathematics and etc.

Collective Mathematical Works

Noah King (Eski)

New Formulas for the Trigonometry of Isosceles
Triangles

Noah G. King (Eski)

518-354-0396

king240@canton.edu

Abstract: This paper discusses a new formula to
solve for the angle of isosceles triangles given
proportion of differentiated unequivalent edge length
versus the always-equivalent edges. The equation is
equal to the arcsine of (X/2) multiplied by two. This
formula is to be called 'isn^-1(x)', short for inverse
isosceles sine equivalent. Also discussed is the non-
inverse 'isn(x)' which is the cyclic function which is
determined to correspond as the equivalent of the
classic sine formula to isosceles triangles. The
formula set is clearly superior and powerful at
calculating the angle and measure of any given obtuse
or acute triangle of an unclassified type, as well as

unifying a simplest-fit formula across all types of triangle.

Equations:

The inverse isosceles sine, with the inequal over one of the equal edge lengths as x:

$$isn^{-1}(x) = \sum_{n=0}^{\infty}\left(\frac{(2n)!x^{2n+1}}{16^n(n!)^2(2n+1)}\right)$$

$$= \sum_{n=0}^{\infty}\left(\frac{(2n)! * x^{(2n+1)}}{((2^{2n} * (n!)^2 * (2n+1)) * 2^{(2n)})}\right)$$

$$\cong \left(\frac{1 * x^{(1)}}{((1*1*1)*1)}\right) + \left(\frac{2 * x^{(3)}}{((4*1*3)*4)}\right) + \left(\frac{24 * x^{(5)}}{((16*4*5)*16)}\right) + \left(\frac{720 * x^{(7)}}{((64*36*7)*64)}\right)...$$

'Isosceles Inverse Cosine' ics^-1(x), to solve the angle of any triangle with three known side lengths:

$$ics^{-1}(x) = \left(\pi - \sum_{n=0}^{\infty}\left(\frac{(2n)!x^{2n+1}}{16^n(n!)^2(2n+1)}\right)\right) =$$

$$\pi - isn^{-1}(x) =$$

$$(\pi - \sum_{n=0}^{\infty}\left(\frac{(2n)! * x^{(2n+1)}}{((2^{2n} * (n!)^2 * (2n+1)) * 2^{(2n)})}\right))$$

$$\cong \pi - \left(\left(\frac{1 * x^{(1)}}{((1*1*1)*1)}\right) + \left(\frac{2 * x^{(3)}}{((4*1*3)*4)}\right) + \left(\frac{24 * x^{(5)}}{((16*4*5)*16)}\right) + \left(\frac{720 * x^{(7)}}{((64*36*7)*64)}\right)...\right)$$

...where in any triangle besides isosceles:

$$x = (\frac{(adjacent\ 1)^2 + (adjacent\ 2)^2 - (opposite)^2}{(adjacent\ 1) * (adjacent\ 2)})$$

For the cyclic function, i= sqrt(-1) and with
isn^-1 angle theta as x:

$$isn(x) = \sum_{n=1}^{\infty} \left(\frac{x^{2n}}{(2n)!} * i^n \right)$$

$$\cong \left(\frac{x^2}{2} - \frac{x^4}{4*3*2*1} + \frac{x^6}{6*5*4*3*2*1} - \frac{x^8}{8*7*6*5*4*3*2*1} \cdots \right)$$

$and : 1 - isn(x)$

$$= \sum_{n=0}^{\infty} \left(\left(\frac{x^{2n}}{(2n)!} \right) * i^{(n-1)} \right)$$

$$\cong \left(1 - \frac{x^2}{2} + \frac{x^4}{4*3*2*1} - \frac{x^6}{6*5*4*3*2*1} + \frac{x^8}{8*7*6*5*4*3*2*1} \cdots \right)$$

The equations appear to exponentiate upon the exact
values of the angular and proportionate measures for
the given notation, though may take large numbers of
iterations to become precise.

Another series of cyclic functions to be defined
by modifications of isn(x) that will calculate and be
equal to to proportion from the radius of the graph or
angle to the respective x and y components on the
cartesian plane is obviously managable from here. They
are to be named isx(x) and isy(x).

$$isx(\gamma) = (1 - isn(\gamma))$$

$$isy(\gamma) = \left(1 - isn\left(\gamma - \left(\frac{\pi}{2} \right) \right) \right)$$

Additional Applications:

In physics, the isn^-1 equation is shown to be used to compute vector combinations and tensors more effectively. It can also be used in pure vector mathematics with ease.

Given right triangles as reference to x and y components plus magnitudes of forces, like so:

Given force or vector set $F(n\ forces)$ and $P's(n\ components)$

of any vector for example:

$$Px = \{Px(n)\}$$
$$Py = \{Py(n)\}$$
$$Pf = \{Pf(n)\}$$

...the resultant force on a point or object will be given the components:

$$\gamma = \left(\left(\frac{Py}{\sqrt{Py^2}}\right) * ics^{-1}\left(\frac{2Px^2}{Px * Pf}\right)\right) or...$$

$$\gamma = \left(isn^{-1}\left(\frac{Py\sqrt{(Pf - Px)^2 + (Py^2)}}{|Py| * Pf}\right)\right) and...$$

$$Fx = (Pf * (isx(\gamma)))$$
$$Fy = (Pf * (isy(\gamma))) \text{ so that...}$$

The force's angle equals:

$$\theta = ics^{-1}\left(\frac{Fx^2 + \sqrt{Fx^2 + Fy^{2^2} - Fy^2}}{Fx * \sqrt{Fx^2 + Fy^{2^2}}}\right)$$

...and its magnitude equals:

$$\sqrt{Fx^2 + Fy^2}$$

This proves that the theorem is effective in calculating dimensional components of vectors in a vastly alternative and simplified method. Positive and negative x values are right at hand in this setup, while y values must be differentiated. This at least preserves the symmetry of the reference dimension and, using corresponding signs for if the angle to a radian is negative, one can obtain full 360 degree coordinates, or adapt the formula to more advanced purposes. Less formulas by far are needed for calculation of geometry and physics with this vector method, and the calculations can be put in one piece with less external operations or operators.

A calculation for the ideal circumference of the perfect circumscription of a known SSS triangle can be done as follows, where variable 'side' varies from the longest to shortest side of the triangle for the calculation of the correspondent circumscription, semiscription, or inscription:

$$x = \left(\frac{(adjacent\ 1)^2 + (adjacent\ 2)^2 - (opposite)^2}{(adjacent\ 1) * (adjacent\ 2)} \right)$$

$$C = \left(\left(\pi + |isn^{-1}(x)| \right) \right) * side$$

Additionally, the ratio of a circle diameter to circumference PI = 3.14159… can be calculated using the infinite series simplified from isn^-1(x):

$$\pi = \sum_{n=0}^{\infty} \left(\frac{(2n)!2^{-2n+1}}{(n!)^2(2n+1)} \right)$$

One more theoretical set of expressions which works best using isn^-1(x) is the set which produces the average of all angles of x->y and x->z, as well as the one with the average of x->y and y->z, in a set of points within an object in equidistant and regular conditions. The average value of the sum pairs of these angles respectively condense unique irrational numbers that vary across a set of possible values based on the scale, rotation and perspective of an object or region in a graph. The single value produced by the set makes it possible to correspond a single number of many decimals to an expansion that reproduces the graph of the object, and in the latter case may be able to locate the object by its center of point sample density when it is part of a larger equation. This method relies only on the average of the sum of the defining angles to a region R at its perspective, the equation and process can be expressed as:

$$Graph(x,y,z) = \sum_{x,y,z=(x_1,y_1,z_1)}^{(x_m,y_m,z_m)} \left(\frac{(isn^\wedge - 1(x,y) + isn^\wedge - 1(x,z))^{\sqrt{x^2+y^2+z^2}}}{m} \right)$$

Where m will represent the number of x,y,z coordinates to fill the region that are used as a 3-dimensional list.

Simply put, this method will return irrational numbers that correspond uniquely and identically to one particular graph with optional transformations.

One other transform to reveal roughly the approximate harmonic signature of a series of samples represented as f(x) in terms of isn(x) frequencies(x) and amplitudes(y), and letting k represent the number of recorded samples of f(x):

$$Harmonics(f(x)) = \sum_{n=0}^{k} \left(\frac{isy(f(n)*x)*\left(\frac{k}{x+n}\right)}{k} \right)$$

Discussion:

The isn, isn^-1 and ics^-1 formulas provide a unifying method whereat it is optimally easy to solve any other type of triangles (in relation to the relative component of the positions or sides of the adjacent points to the angle). The sin and arcsin methods are specific to components and do not provide method to solve other triangles in a way where the

modifications are significant to accessible properties
of the triangles.

Since no other formula has the proper blend of
consistency, reflexive clarity and ease with
calculating all types of triangles, and this formula
is computationally superior for obtaining transformed
physical predictions and simulations as well as data
on geometric systems, ics^-1 ought to be used as a
unifying trigonometric equation in the process of
relating expressions and triangles to each other or
their other measurements. it may very well be the
superior method of calculating any type of triangle
from an educational and computational perspective, as
well as from a standpoint in search of algebraic
abstraction of geometric properties.

In ultimatum:

The expressional components of 'ics^-1 ready'
triangular conversion coefficients for proper angle
calculation are more universally supported by this
formula than the cos^-1 theorem can rationally express
to, making it computationally superior for closed
systems and computing large networks of geometrically
or mathematically bound variable systems that would
correspond to everyday applied mathematics. Using a
simple script and algebraic rules every geometric and
physical formulation may be given a uniform
distribution and ease of reference between different
theorems, potentially simplifying together series of
expressions in systems that are closed or of high
complexity, not easily possible with sin or sin^-1 on
an input or any scale of algebraically easily
transformable level. It is suggested that many
computational systems and graphing calculator make an
easy use of this function or its generated dataset and
its conversion utilities for a geometric, spatial or
coordinate network.

<u>Data Availability Statement, and Conflicts of Interest</u>:

The data and calculations generated during and/or analysed during this study are not included in this paper. The calculations are simple and are easily tested for proof that they converge and are equivalent to the arcsine of (X/2) multiplied by two. The example data is not included for the simple reason that it does not help demonstrate any further topics than is discussed so far in the paper. Due to the fact that any standard calculator or convergence test will yield the results which are claimed in this text, they are available from the corresponding author on reasonable request.

On behalf of all authors, the corresponding author states that there is no conflict of interest.

References:

1. Wolfram|Alpha. Calculators and Convergence Testing Applets; Information on Modern Trigonometric Formulas.
   https://www.wolframalpha.com/
2. Convergence test
   https://www.wolframalpha.com/widgets/view.jsp?id=8e68e1789f7b8fbbbb1a44197d369ad1.
3. Inverse sine
   https://mathworld.wolfram.com/InverseSine.html.
4. Law of cosines
   https://mathworld.wolfram.com/LawofCosines.html.
5. PI-Wolfram language documentation
   https://reference.wolfram.com/language/ref/Pi.html.

Postulate of Time as an Imaginary Dimension(s), Physical Space as Real Dimensions, and the Behavior of Negative versus Positive Numbers

Noah G. King (Eski)

518-354-0396

king240@canton.edu

Postulate: Imaginary dimensions are correspondent to time or chronological phenomena, where mathematical points and data are the conceptual real physical architecture encased, to be used to model reality as experienced. Additionally, negative numbers have long been thought to multiply to form positives, but mathematics is more easy as a whole when negative values always multiply to negatives, and are equivalent to positives save that the product between a negative and a positive is imaginary and thus both positive and negative.

Evidence:

Point and function derivatives and integrals are logically symmetrical in/ex – trusive analysis of points on a graph. Logical series coordinate through time in a complex manner. Physically, it is deterministic that a major degree of time is to provide fragmented dimensional components with real and imaginary parts against each other or in systems in different parts of reality. Time is always the imaginary part of these functions, spatial dimensions are real and non-conformational, though time is an overarching container for these expressions.

In addition, negative numbers have long been thought to multiply to form positives, but mathematics is more easy as a whole when negative values always multiply to negatives, and are equivalent to positives save that the product between a negative and a positive is imaginary and thus both positive and negative. This can easily be tested and this methodistic procedure **simply requires less assumptions,** while retaining the simplified functionality of all major applied mathematics functions and formulas. That is to also imply that:

1. +/- symbols as operators are directionally biased to add or subtract from the relevant direction of the handedness of graphs, so that adding two negatives produces further negatives, while subtracting a positive from a negative will add its value back towards/through zero.

2. Exponential graphs and radical graphs are to be considered for in that the signed power/root is simply going to become ambiguous when different, but treated as if it were as we currently see a positive root or power to a positive number, and the result is always the same, positive or negative, with the exception that the result is indeterminate and imaginary(positive and negative) when the signs are on either hand, different.

Then, we can derive theoretically a variety of useful logical workarounds to achieve some of the other graphs we were more familiar with using the old style of numeric negativity:

a. (-1/(-x)) will result in the value -1/X, (+1/(+x)) will result in the value 1/X, but if the signs are different they will result in the consequent proper-sign (+/-)X/1, and the net absolute can be taken to result in graphs such as the continued 2^x, in the expression now absolute |-1/(2^x)|.

b. Multiplying a graph by (+/-)1 can result in the negative or positive side of the graph flipping its orientation while the other stays the same.

c. This opens up so that we can assume 0*0 is
   equal to 1 as well as 0/0 can equal 1,
   which fixes a hole in the logic behind
   factorials such as factorial 0=1. The
   factorization of zero is infinite but also
   empty but can equillibriate logically to
   one.

See the logical correspondence of any function, sequence or
logical construct as a whole to be continuous of a logical
degree, and that an imaginary number – rooted reflection of
the trend to be identical to only a possible and self-
sustaining neutral symmetrical system, or else a real
logical cyclic phenomena. This is aligned to the structure
and behavior of matter, or the logical order of any system
of mathematics with sub or consequent derivatives (or
removals from) or integrals (or additions to) of or without
any mathematical systems, or any modifications thereat,
especially those designed to model real systems we
currently perceive.

The scalar behavior of numbers when their values are infinite

Noah G. King (Eski)

518-354-0396

king240@canton.edu

Postulate: *The assumption is that because decimals terminate at what is infinitely small and they represent the set of possible integers, the correspondent scale of numerators to the denominator of infinity in increments is digital and no longer accounts for decimals. That said, an equational graph with a numeric method which equates the rounded value of X is used and the integrals and areas from 0-x of the graphs is to be postulated, with the proportion of adjacent slopes, to be the related to the relative geometry of infinitely large values inverses with respect to their multiples and regular numbers as they are physically measured. While computers and calculators may express the value of (x*(inf))/(inf) as equals to y=x, it is just as simple to equate infinite values using this relationship and derive potential statistically accurate substitute values in typical algebraic limits of undefined points. The next section includes proof.*

Equations and proof:

Because :

$$\left(\frac{x}{k}\right)^{\infty} = \left(\frac{x}{k}\right)\infty = \infty \ if \ k < x; \frac{1}{\left(\frac{x}{k}\right)\infty} \ if \ k > x; 1 \ if \ k = x \ldots$$

This means that:

$$infscale(x) = \sum_{k=1}^{\infty} \frac{\left(\frac{x}{k}\right)^{\infty}}{\infty^{\left(\frac{x}{k}\right)}}$$

$$= \sum_{k=1}^{\infty} \frac{\left(\frac{x}{k}\right)^{\infty}}{\infty * \left(\frac{x}{k}\right)}$$

$infscale(x)$ also equals:

$$= \sum_{k=1}^{\infty} \frac{\left(\frac{x}{k}\right)^{2\infty}}{(2 * \infty)^{\left(\frac{x}{k}\right)}}$$

$$= \sum_{k=1}^{\infty} \frac{2x \left(\frac{k}{x}\right)^{\infty}}{\infty^{\left(\frac{k}{x}\right)}}$$

Infscale(x) can be presumed to equal

the scalar value of

x as a multiple of infinity if $1 = \infty$ *because...*

If the operators of infinity in this

equation are a finite number,

the equation functions to produce

repetitive infinite sums

converging from equalling to the formula

$y = x$ *when* $\infty = 1$

and as $1 = \infty$ *we see it becomes*

the rounded value at x.

The equation technically converges with high precision when the sigma iterations, the power of the numerator and the base of the denominator are all adjusted towards inf. exactly from infscale([-1<x<1]); but since infinity has not been quantified and its power principle of itself as a base is multiplied by an infinite power, it is theoretically an always-infinite power even when inf. is put to the first power; since infinites may be an undefined set whereat the relation 1*inf != infinity because 1 is technically 1/inf to infinity. This implies we can safely use computational space to refer to infinity as a relationally infinitely scaled set of possible numbers and thereby it holds relational properties whereat infinitely small decimals may count up as integers when scaled linearly by inf.

Because any possible numeric substitution for inf. as a power to a number is quantifyably a decimal to infinity as well, numerically we can relate powers of infinity to the integer-power infinite roots analogy and vice versa, and find that only scalar proportions n/inf. can be multiplied by inf. to achieve a realistic numeric value that is finitely expressible. Therefore without implying infinite zeros after ever decimal notation, you cannot effciently imply sub-decimals(i.e. 3.5.2) as a true number as the process is redundant.

In mathematics the scalar identities of sub-decimals would be consequent divisions of infinity; and when used in operations with infinity can yield new numeric methods for the potential geometric establishment of new number construction processes for infinitely large or small numbers, and expanding them to integer analogies later. In the geometric theory this is to imply infinity as a transcendental number that still posesses geometric properties on our numbers when certain rules are

applied to arbitrary calculations that
use it, since inf. is not previously able
to be expressed or used as a true number
value in computers. It would also be the
only number not allow for algebraic
simplification.

This allows us to conclude that scale(x)
is the representation of all possible
numbers when $x = \infty x$. This is probably
because decimals are infinitely small
when compared to their whole values when
x is divided by infinity, or when 0-1 is
expanded as the range of possible
integers by their order of magnitude.

Technologically, numbers from the
derivatives or integrals or other
transformations of this formula may
reference to significant numbers in
certain spatial or transformational sets,
which can be assistive or relevant in
geometrical problems, especially when
given to simulate or emulate infinite
variables.

Study on Novel Geometric and Numeric Methods

Noah G. King (Eski)

518-354-0396

king240@canton.edu

Abstract: *This paper discusses a variety of newly discovered numbers that have unique properties in algebra that may have interesting applications in geometric and mathematical problem-solving, as well as exhibit plentiful beauty in their self-structured relational patterns.*

Introduction:

Numbers that uniquely possess logically simple algebraic properties that pertain exclusively to themselves are discussed in this paper. The properties are expressible as simple logical equalities where a transformation of the number involved in a simple or common way results in a concurrent transformation of the same number by an alternate route, or in general a relationship of the number to itself that exhibits an increased self-reference when compared against other numbers.

Here is the list of numbers and their values, along with their defining properties which make them special found in the next section:

T = 1.8795…

```
L  =   1.8393…

P  =   1.813…

S  =   1.7768…

D  =   1.7549…

J  =   1.7569…

I  =   1.7286…

O  =   1.7221…

o  =   1.6938…

H  =   1.678…

X  =   1.6603…

Q  =   1.63960…

E  =   1.5652…

Z  =   1.5549…

K  =   1.5204…

N  =   1.4966…

Y  =   1.4656…

W  =   1.4534…

F  =   1.44138…

r  =   1.4364…

f  =  1.4343…

B  =   1.38152…
```

A = 1.3660…

U = 1.3392…

M = 1.1975807343…

C = 1.2488…

G = 1.1993…

m = 1.19743…

R = 0.4759…

V = 0.4360…

n = 0.2324…

Findings:

These numbers discussed in this paper, referenced in the above section, have a variety of properties which are likely to help to formulate various mathematical equations in alternate ways. Their properties are of intense mathematical beauty in the opinion of your author.

Noted, some of these identities have strange and potentially useful properties – and, despite being hard to calculate by hand, they have relations that make them easy to find and exact with just a calculator. In various numeric systems, it becomes easy to wonder if these numbers can have implementation in shortening long calculative processes through logical 'loopholes,' or substituting expressions with ones of their own archetype. They can also make perfect geometric patterns or hold a root in geometric problem solving. One good example is Atta A, a number that equals $(1/(sqrt(3)-1))$ and generates a perfect scaling rectangular tessellation:

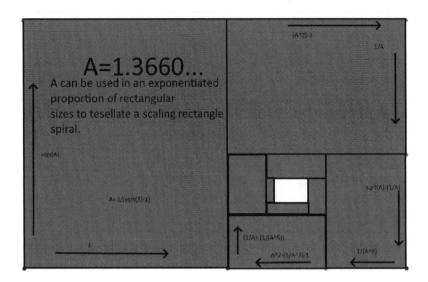

Another example is the spiral for Metta M, which spirals through a series of rectangles scaled by the first, third, fifth and so on rectangles:

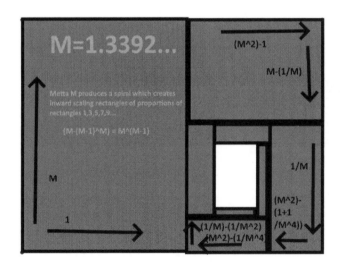

Another example is the spiral for Setta S, which spirals through a series of alternatedly arranged logical statements for its side lengths while maintaining a repeating order of vertices for rectangles in a tessellation:

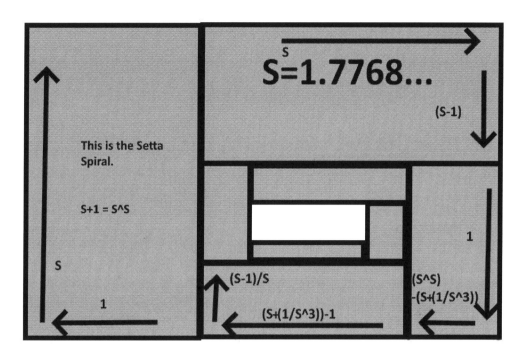

This is the Setta Spiral.

$S+1 = S\text{^}S$

$S = 1.7768...$

Another example is the spiral for Detta D, a number with a plethora of properties, which spirals through a perfect square:

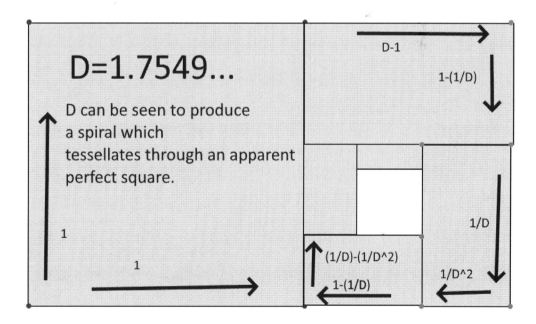

$D = 1.7549...$

D can be seen to produce a spiral which tessellates through an apparent perfect square.

One more good example is Letta L, a number (not first discovered by the author, known collectively as the Tribonacci Constant) that, with its inverse is a root, or the zero-value of the equation of $((X^2-X)/(X-1))-X$ and generates yet another perfect scaling rectangular tessellation:

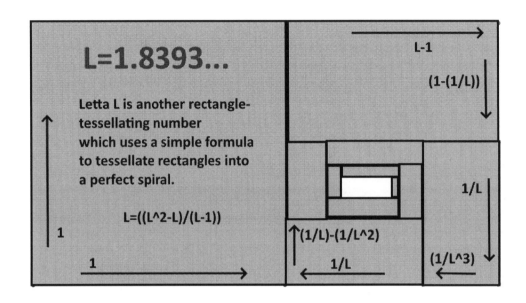

This paper will detail to you the values and properties of each of the respective numbers. They may be subject to be used in the calculation of natural progressions, processes, and geometrical relationships. These numbers may possess properties that can, much like words in a language, simplify modern mathematical operations used in conjunction, by offering substitutions and simplifications and alternate constant-based methods to express other forms of logical transformations of numbers. More effort is required in investigating into the functions of these numbers, but their general importance and reasons for consideration are evident in the properties of the numbers herein listed, expressing the properties which can simplify or substitute for mathematical transformations where they are harder to write into an equation: The properties of the previously listed numbers will now be explained on the next page

Let number Meum M satisfy the equation:

With two other numbers as additional solutions:

$(2^M)/(M^2) - (M-1)/M = M^2$

$((M-1)*M + (M-1)*(1/M)) = ((2^M)/(M^2) - M)$

$(((2^M)/(M^2)) + 1) = ((M^3) + M))$

$(((2^M)/(M^2)) - M^3) = (M - 1)$

A very closely calculated number that is within a ten thousandth or two of Meum holds:

$(M-1)^M + (M-1)^(1/M) = (2^M)/(M^2) - M$

Let number Etta E satisfy the equation:

$(1-(E-1))-(1/E) = E-(1/(E-1))$

Let number Atta A satisfy the equation:

$(A+1)/A = A+(A-1)$

$(1/A)-A = (1-(A-1)$

Let number Ketta K satisfy the equation:

$K^K = K+(K-1)^K$

Let number Utta U satisfy the equation:

$U-(U-1)^U = U^(U-1)$

Let number Wetta W satisfy the equation:

$1+(W-1)^2 = sqrt(W) = 1+((W-1)/(sqrt(W)+1))$

Let number Zetta Z satisfy the equation:

$Z-(Z-1)^2 = sqrt(Z)$

Let number Fetta F satisfy the equation:

F = (F+1)/(F^F)

Let number Yetta Y satisfy the equation:

Y^2 = 1/(Y-1)

Let number Tetta T satisfy the equation:

T^T = (T+1)/(T-1)

Let number Metta m satisfy the equation:

((m^2)/2)+1 = m^3

Let number Detta D satisfy the equation:

(1/(D-1)) = 2*(D-1)

(1/(D-1))-1 = 1/D^2

(D-1)^2 = 1/D

D^2 = D+(1/(D-1))

D = (1/(D-1))^2

D+1 = D*(1+(D-1)^2)

Let number Itta I satisfy the equation:

D+1 = D*(1+(D-1)^2)

Let number Cetta C satisfy the equation:

(1+(C-1)^2) = (C/(1+(C-1)^C))),

(1+(C-1)^C) = (C/(1+(C-1)^2)))

Let number Retta R satisfy the equation:

(R^R)/R = R+1

Let number Vetta V satisfy the equation:

V/(V^(V+1)) = V+1

Let number Rutta r satisfy the equation:

1/r = 1-(r-1)^r

Let number Hetta H satisfy the equation:

(2^H)/(H^2) + H = H^2

Let number Xetta X satisfy the equation:

(1/X) = 1-((X^(X-1))-1)

Let number Letta L satisfy the equation:

L^2 = (L+1)/(L-1)

Let number Getta G satisfy the equation:

((G-1)^G)/(G-1) = 1-((G-1)/((G-1)^(G-1)))

Let number Petta P satisfy the equation:

(P^(P-1)-(P-1))*(1-(P-1))*P = sqrt(P)

Let number Setta S satisfy the equation:

S+1 = S^S

Let number Jetta J satisfy the equation:

J^J = (J-1)^(1-(J-1))

Let number Otta O satisfy the equation:

O^O - (O-1)^(1/O) = O

Let number Outta o satisfy the equation:

o/(o-1) = o^o

(o^2/(o-1))-o^o = o

Let number Netta N satisfy the equation:

N^N - ((N-1)/N) = N

Let number Futta f and Nutta n satisfy the
equation:

Both are roots of the equation:

(1/(1-(1-(1/X))))-(1-(X-1)) = (1+(1-(1/X))-(X-1))

They also relate by:

(1-(1/f))+1 = 1/(1-n)

(1/(1-n))-1 = 1-1/f

1/(1-(f-1)) = 1+(1/(1+(1-(1/f))))

Let number Quetta Q and Betta B satisfy the
equations:

1-(B-(1/B))=(B^B)-1

B^B-(1-(B-1) = 1/B

Q^Q - 1/Q = Q

References: _No references or sources on these numbers or the information on this paper were found, perused, or used, and the work done and the theory involved was developed wholly by the author using their own background of knowledge and calculative methods._

Formulations and Expressions of Two Geometrical Point Connection Complexity Graph Series Sequences

Noah G. King (Eski)

518-354-0396

king240@canton.edu

Description: *The number sequence to enumerate the possibilites of the numeric index of number of vertices in connection in a series of unique graphs, by possible combinations as well as possible combinations with respect to geometric disorganization or hierarchy, shall now be listed following the governing functions.*

f(x)=2^(sum(i=1->index,i-1))

g(x)=(index)^(sum(i=1->index,i-1))

Software outputs enumerating this sequence:

the number to enumerate the sequence to: 15

index: 1

point pair permutation single connection: 1

index with all possible intersections considered: 1

index: 2

point pair permutation single connection: 2

 index with all possible intersections considered: 2

index: 3

point pair permutation single connection: 8

 index with all possible intersections considered: 27

index: 4

point pair permutation single connection: 64

 index with all possible intersections considered:
4096

index: 5

point pair permutation single connection: 1024

index with all possible intersections considered:
9765625

index: 6

point pair permutation single connection: 32768

 index with all possible intersections considered:
4.7018e+11

index: 7

point pair permutation single connection: 2097152

 index with all possible intersections considered:
5.5855e+17

index: 8

point pair permutation single connection: 268435456

 index with all possible intersections considered:
1.9343e+25

index: 9

point pair permutation single connection: 6.8719e+10

index with all possible intersections considered: 2.2528e+34

index: 10

point pair permutation single connection: 3.5184e+13

index with all possible intersections considered: 1.0000e+45

index: 11

point pair permutation single connection: 3.6029e+16

index with all possible intersections considered: 1.8906e+57

index: 12

point pair permutation single connection: 7.3787e+19

index with all possible intersections considered:
1.6825e+71

index: 13

point pair permutation single connection: 3.0223e+23

index with all possible intersections considered:
7.7194e+86

index: 14

point pair permutation single connection: 2.4759e+27

index with all possible intersections considered:
1.9845e+104

index: 15

point pair permutation single connection: 4.0565e+31

index with all possible intersections considered:
3.0873e+12

Analysis of the Number of Unique Numbers Generated

by Any of Two Identical Number Sets {0-X} Compounded in

Any of the Six Major Mathematical Operations

(powers, roots, multiplication, division, addition, and

subtraction) as Counting Sequences

Noah G. King

518-354-0396

king240@canton.edu

ABSTRACT: This paper discusses how the analysis of the number of unique (counted one for all number identities with identical cosequentials) numbers as a result of the combination of any two numbers up to index can be found to follow divergently asymmetrical sequences in terms of the complexity of the operation to increase or decrease in value of the numbers.

FINDINGS:

The article expresses, in order: the sequences, from most complex decreasing operation to most complex increasing operation. This article now discloses the first 64 values of the sequences:

-Roots sequence:

1, 3, 5, 9, 13, 21, 30, 42, 54, 65, 81, 101, 120, 144, 168, 195, 215, 247, 277, 313, 348, 387, 426, 470, 512, 548, 594, 636, 686, 742, 794, 854, 906, 968, 1030, 1097, 1143, 1215, 1284, 1358, 1430, 1510, 1585, 1669, 1750, 1835, 1920, 2012, 2099, 2171, 2262, 2360, 2456, 2560, 2658, 2765, 2869, 2979, 3087, 3203, 3312, 3432, 3548, 3670…

-Quotient sequence:

1, 3, 5, 9, 13, 21, 25, 37, 45, 57, 65, 85, 93, 117, 129, 145, 161, 193, 205, 241, 257, 281, 301, 345, 361, 401, 425, 461, 485, 541, 557, 617, 649, 689, 721, 769, 793, 865, 901, 949, 981, 1061, 1085, 1169, 1209, 1257, 1301, 1393, 1425, 1509, 1549, 1613, 1661, 1765, 1801, 1881, 1929, 2001, 2057, 2173, 2205, 2325, 2385, 2457…

-Difference sequence:

1, 3, 5, 7, 9, 11, 13, 15, 17, 19, 21, 23, 25, 27, 29, 31, 33, 35, 37, 39, 41, 43, 45, 47, 49, 51, 53, 55, 57, 59, 61, 63, 65, 67, 69, 71, 73, 75, 77, 79, 81, 83, 85, 87, 89, 91, 93, 95, 97, 99, 101, 103, 105, 107, 109, 111, 113, 115, 117, 119, 121, 123, 125, 127…

-Addend sequence:

1, 3, 5, 7, 9, 11, 13, 15, 17, 19, 21, 23, 25, 27, 29, 31, 33, 35, 37, 39, 41, 43, 45, 47, 49, 51, 53, 55, 57, 59, 61, 63, 65, 67, 69, 71, 73, 75, 77, 79, 81, 83, 85, 87, 89, 91, 93, 95, 97, 99, 101, 103, 105, 107, 109, 111, 113, 115, 117, 119, 121, 123, 125, 127…

-Product sequence:

1, 2, 4, 7, 10, 15, 19, 26, 31, 37, 43, 54, 60, 73, 81, 90, 98, 115, 124, 143, 153, 165, 177, 200, 210, 226, 240, 255, 268, 297, 309, 340, 355, 373, 391, 411, 424, 461, 481, 502, 518, 559, 576, 619, 639, 660, 684, 731, 748, 779, 801, 828, 851, 904, 926, 957, 979, 1009, 1039, 1098, 1117, 1178, 1210, 1238…

-Powers sequence:

1, 2, 4, 8, 12, 20, 29, 41, 51, 61, 77, 97, 116, 140, 164, 190, 208, 240, 271, 307, 341, 379, 418, 462, 504, 540, 586, 622, 671, 727, 780, 840, 882, 942, 1004, 1068, 1114, 1186, 1255, 1327, 1398, 1478, 1554, 1638, 1718, 1800, 1885, 1977, 2064, 2136, 2226, 2322, 2417, 2521, 2620, 2724, 2827, 2935, 3043, 3159, 3268, 3388, 3504, 3624…

These are clearly interesting and expansive sequences. It may be of significance that the addition and subtraction sequences are the same and expressed or somehow satisfied by three simple functions, one of which is Y=|2x+1|. It may also be of interest that multiplication and powers sequences diverge at the fourth unit and do not reconverge. Stranger yet, the roots and division sequences are seen to diverge at their seventh values. Two of these, the addition and subtraction sequences are linear, and the others are not of a standard and simply derived curvature.

Analysis of the Sequence of Minimum but Ideal Number of Axioms Generated By Geometric Systems of N Complexity

Noah G. King

518-354-0396

king240@canton.edu

ABSTRACT: This paper details the list of axioms, along with their total number of axioms in sequence, that will always apply and be relevant to geometrical systems of N complexity, or N number of points. It relies on the following:

a) *the points are not always organized regularly or are in any given position*

b) *the points may be congruent or in the same reference but are not identical.*

FINDINGS:

The findings of the sequence to date are as follows:

System Complexity	System Axioms	Value of Index
0		0
1	-Point equals itself -Point does not equal what is not that point -Point may form a regular solid within n-1 dimensions.	3
2	-Point equals itself	5

	-Point does not equal what is not that point -Point forms a regular solid within n-1 dimensions. -Points may be compared -Points may be connected	
3	-Point equals itself -Point does not equal what is not that point -Point forms a regular solid within n-1 dimensions. -Points may be compared -Points may be compared relatively to another -Points may be connected cyclically -Points may exist between or within solids in n-2 dimensions -Point connections may be intersected	8
4	-Point equals itself -Point does not equal what is not that point -Point forms a regular solid within n-1 dimensions. -Points may be compared -Points may be compared relatively to another -Points may be connected	9

	cyclically	
	-Points may exist between or within solids in n-2 dimensions	
	-Point connections may be intersected	
	-Point comparisons may be compared to unconnected comparisons	
5	-Point equals itself	10
	-Point does not equal what is not that point	
	-Point forms a regular solid within n-1 dimensions.	
	-Points may be compared	
	-Points may be compared relatively to another	
	-Points may be connected cyclically	
	-Points may exist between or within solids in n-2 dimensions	
	-Point connections may intersect	
	-Point comparisons may be compared to unconnected comparisons	
	-Point comparison chains may be compared relatively to each other.	
6	-Point equals itself	11
	-Point does not equal what is not that point	
	-Point forms a regular solid	

	within n-1 dimensions.	
	-Points may be compared	
	-Points may be compared relatively to another	
	-Points may be connected cyclically	
	-Points may exist between or within solids in n-2 dimensions	
	-Point connections may intersect	
	-Point comparisons may be compared to unconnected comparisons	
	-Point comparison chains may be compared relatively to each other.	
	-Point comparison chains may be compared to each other.	

CONCLUSION:

The findings of the sequence to date are as follows:

The most reasonable number of encompassing statements in a randomly assigned geometric system for n complexity is 0-6 is 0,3,5,8,9,10,11 for that number of points in a geometric system.

BIODES AND TRANSRATIOMETERS

NOAH KING

king240@canton.edu

5183540396

ABSTRACT.   This paper discusses how the introduction of 'Biodes' and 'Transratiometers' could potentially change how computer chips are made. These two new components reduce some simple processes in size drastically, and allow for greater fidelity in current, as well as recycling current more efficiently in some circuits. The Biode is a single-input, two output diode that allows a single signal to be split into two signals. The Transratiometer is a two-input, two-output transistor that functions individually as its own logic gate, and can also be used to proportionalize or split current output values. These devices help to better recycle current and simplify circuits in computer chips and transistor chips, using less wattage.

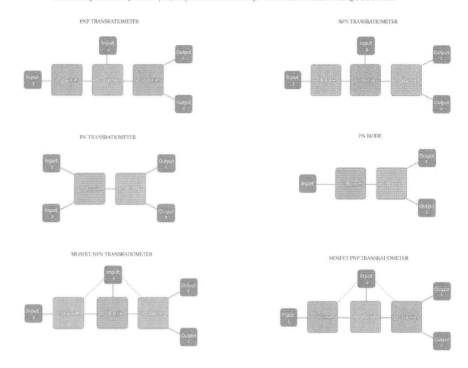

BIODES AND TRANSRATIOMETERS BLOCK DIAGRAM

The Biode and Transratiometer are two logical processing devices that can simplify, make more efficient, and lessen the size of modern computing circuits. They are CMOS compatible and will be designed to prevent current sharing and thermal runaway (Or at least have increased emmitter resistance to prevent current sharing and still save power and space). They function better than multiple-emitter transistors in classic TTL logic, and are faster.

Biodes and Transratiometers and Their Function in Modern Circuits

Modern computing architecture is subdivided into logical processing components called logic gates, common to silicon chips and other computational mechanisms. They often consist of transistors, resistors, capacitors and other small components that can be used to make complex computations, and are the basis of modern computational technology.

How they work is essentially based on the directions of current, placement of values, and the many various logic gates that decide how that current is handled, and what signals are computed, how they interact with resistance, and in conjunction. Reversible computing and a few other types of data interpretation change the way data is ran, but the computing relies on the same basic principles.

When transistors overfill and release power, they put out more energy than what's put onto one terminal, more than is necessary for a signal, and often resistor pairs or other types of power recycling methods are used to compensate for that. However, the biode and transratiometer simplify the chip process greatly.

The biode is a two output diode, easily sketched by wafer etching machines as basically a reverse-in-current transistor. What it does, is serves to reduce the current of input and irreversibly and non-interchangeably give two distinct currents to two lines from the output of a transistor.

This is a more effective way to recycle energy from transistors, but not only that; the biode is capable of sending duplicate signals and interlinking simultaneous computing processes together, to check many logical samples from one output at once against the output, or interrelate number values together on the chips.

The transratiometer is basically a two-output transistor, capable of shedding half or a ratio of its current instantaneously in a chip without the use of two transistors, and effectively works as a single component AND gate.

Proportions of current can be serialized through the ratio of conduction from either contact to the two outputs, so data can be processed linearly through the transratiometer as well. It effectively simplifies single-transistor outputs into regular currents that are normal to the system as well as possessing the ability to divert serial/output streams.

The biode may replace the resistor and eliminate the function of the transratiometer in the process of fixing currents for long serialized streams or numbers of transistors, but the transratiometer eliminates the need for resistors and diodes and can coordinate complex current levels in multiple directions for the serial streams, as well as make XOR and NOT gates much more efficient spatially and in wattage.

Because this uses less space than the models with resistors, diodes, and more transistors, and also because power can be stacked into biodes and split with transratiometers, the biode and transratiometer are good at operating after, or in the case of the transratiometer, in place of transistor outputs in AND, NOR, XOR gates, as well as complex systems such as adders and subtractors, and addressors and registers.

Key

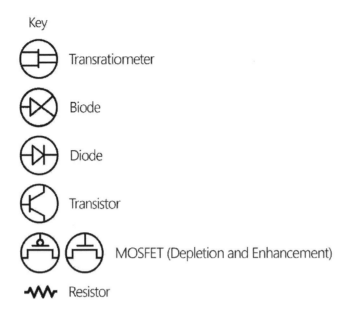

Transratiometer

Biode

Diode

Transistor

MOSFET (Depletion and Enhancement)

-\/\/\- Resistor

AND Gates

Fig. 1

Fig. 2

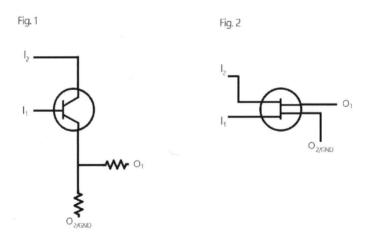

Fig. 1 Represents a traditional AND gate, a total

of 3 components. Fig. 2 Represents an AND gate
utilizing the Transratiometer component, for 1
component in total. The first circuit has a higher
current than I(2) or I(1) for output, necessitating
the resistor, the second circuit has a mean of the two
current inputs as outputs.

OR Gates

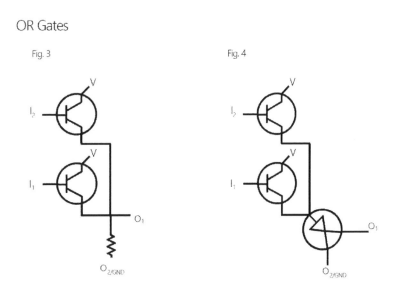

Fig. 3   Fig. 4

Fig. 3 Represents a tradtional OR gate, using a
resistor due to excess wattage at the output. Fig. 4
Represents a similar OR gate using a biode component,
reducing overall wattage.

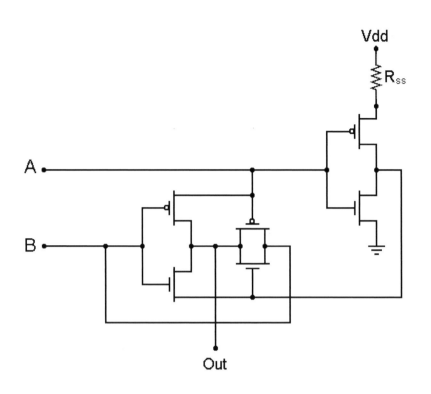

[https://en.wikipedia.org/wiki/XOR_gate] Image depicting an XOR gate, for a total of 7 components.

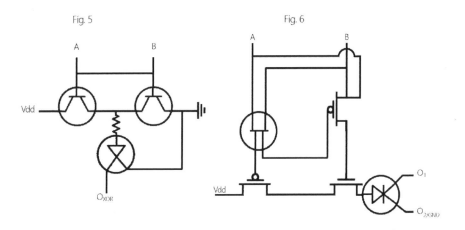

Fig. 5 and Fig. 6 represent XOR Gates that use less components and perform virtually the same function, Fig. 5 using a Biode, and Fig. 6 using a Transratiometer.

.

NOT Gates

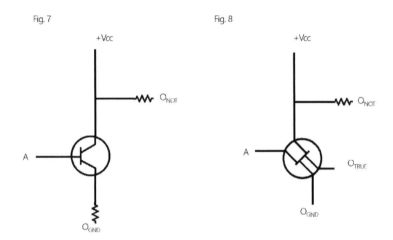

Fig. 7

Fig. 8

Fig.6 Represents a traditional NOT gate. Fig.7 Represents a NOT gate using a transratiometer, which can also be used to supply an additional TRUE signal, and does not require an additional resistor to ground.

.

NAND Gate

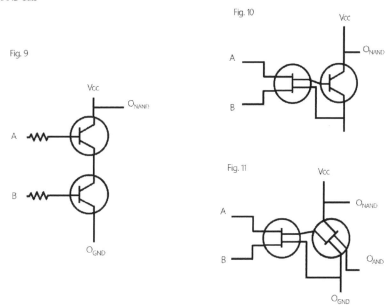

Fig. 9

Fig. 10

Fig. 11

Fig. 8 Represents a traditional NAND gate. Fig. 9 Represents a NAND gate using a transratiometer, which splits the total value of the combined inputs A and B, removing the need for resistors. Fig. 10 represents a similar circuit where the transistor component is replaced with a transratiometer, allowing the gate to save an additional AND value.

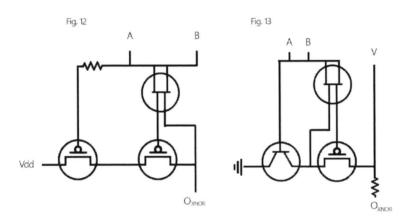

Fig. 12

Fig. 13

Fig. 12 and Fig. 13 represent 2 XNOR gates, showing how transratiometers can be used in different ways as an AND signal component to solve a circuit. Here are figures 13, 14:

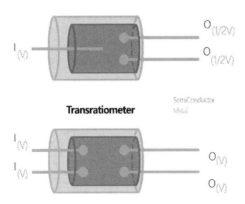

Biode

Transratiometer

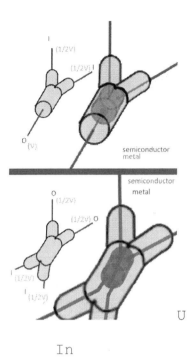

I (1/2V)

(1/2V) I

O (V)

semiconductor
metal

semiconductor
metal

O (1/2V)

(1/2V) O

i (1/2V)

I (1/2V)

U

In

13, left, and 14 at right, serves to demonstrate a possible construction of biodes and transratiometers, which with an analogously designed one-output, two-input transistor, can potentially be made from Electrum, a theoretical ~16.5% Boron to ~83.5% Silicon molar ratio alloy. It is suggested this P-doped semiconductor alloy may have a electrical behavior that lets current through at thresholds by kinetic disequilibria across a single semiconductor junction.

Alternatively, this page now contains basic diagrams of more traditionally designed biodes and transratiometers:

BIODES AND TRANSRATIOMETERS

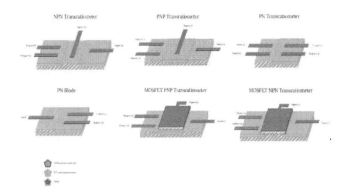

Made in the USA
Monee, IL
30 May 2022

e2dbf583-eed3-4bb8-8807-42ae2aa29c5cR02